HIV/AIDS

HOW EXPERTS SPREAD THE VIRUS

(AN EYE – OPENER)

USHONYE AYIM

*FOREWORD BY A
SEASONED MEDICAL
PERSONNEL*

<u>ONE WORD</u>

"SCIENCE

IN ORDER TO BE USEFUL

MUST

EXCHANGE ITS AUTHORITY

FOR A

CREATIVITY".

- CEES HAMELINK

"WHERE IS THE CONSCIENCE OF SCIENCE

IF EXPERTS COULD SPREAD THE VIRUS?"

- USHONYE AYIM

HIV/AIDS

HOW EXPERTS

SPREAD

THE VIRUS

(AN EYE – OPENER)

USHONYE AYIM

ISBN 978-1-4116-4446-5

Published simultaneously
In the United States and in Nigeria
By

Lulu Press Inc.
3101 Hillsborough Street
Raleigh
NC27607
North Carolina
USA.

Nkymbe Evolutionary Books
3, Adeshi Ntaji Street
Ishiborr
Ogoja, Nigeria.

CONTENTS

DEDICATION

To

MY DEAREST SISTER

MRS. AGBO ALICE ADINYA (NEE AYIM)

FOREWORD

Several times in the past death has visited mankind with deadly diseases that have threatened it. Several diseases – tuberculosis and leprosy, malaria, chicken pox, cholera, lassa fever, typhoid fever and others have all taken their turns in the battle against mankind. Most often than not mankind has triumphed. Today, it is the Human Immuno-deficiency Virus (HIV) Acquired Immune Deficiency Syndrome (AIDS). This has dealt such a devastating blow on mankind so much so that mankind has lost valuable sleep since it was first identified. An unimaginable amount of resources – material and intellectual, has been committed to the quest to unearth the mystery surrounding this 'angel of death'. So much has been achieved! A lot is now known about the virus, its incubation period, mode of transmission etc. Even though so much has been discovered, the battle has been far from over. All these attempts have only yielded preventive and

Palliative measures. A cure, elusive! Everybody, including those in the health profession, scientists and religious Community, is agitated. People in all parts of the world and indeed all walks of life have dedicated their lives and resources to ensure that a cure is found for this pandemic. It is in this same spirit that Evangelist Ushonye Ayim seeks to make his contribution in this Ten Chapter Volume. He has tried to point out where there seem to be curves in the way towards achieving success in the battle against the scourge. In doing so he has quoted the works of experts and placed these side by side with his "Criticisms' this is by no means intended to ridicule, insult them or rubbish their work, but it is intended to elicit more explanation where his observations score a point, galvanize thinkers to think deeper, researchers and scientists to expand their horizon, and may be, give some direction to further improve the arsenal that would bring the battle against HIV/AIDS to a victorious conclusion.

The reader is therefore encouraged to read this volume with the highest sense of responsibility and objectivity. This battle is for all of us and the battle must be won. Enjoy your reading.

Inyambe, Matthew A.
Principal Nursing Officer
Lutheran Hospital Yahe – Yala
Cross River State
Nigeria.

THANK YOU

I do not hesitate to confess I remain aggressively grateful to the Almighty God for the strength and power in reaching the discovery uncovered in this piece of writing.

Also indebted to is Mrs. Ayum Sarah Asanga and Rev. Sr. (Dr.) Kate Okpa, for the prompt dispatch to me of valuable source materials, without which this work could not have been successful. To you Mr. James Alicha (JAMEL), Mr. Clement Inyambe, Mr. Chris Inyang, Mr. Sam Egbala, Mr. Vitalis Ugoh, Mr. Ogbang Akwaji, Mr. Peter U. Inyambe, Mr. Simon Iyaji, Mr. Emeka Umejesi, Mr. Peter Ishone Adi, Evang. Asho Inyambe, Mr. Nnoyi Igri Okon and Mr. Mesembe Ita Edet, I thank you for your incomparable encouragement since I came into permanent contact with you.

Worthy of appreciation is Senator Greg. I. Ngaji, Hon. Vena Ikem, Mr. Joe Eweh, Mr. Gabe Onah, Dr. (Mrs.) Nchajeno Owan, Umenekor Iyaji,

Mr. Willy Ogon, Bro. Abba Ubaka, Mr. Emenike Ezeigwe, Dr. Bassey Igri Okon and Dr. Sunny Ezigbo. To you I confess my complete indebtedness.

Thank you Mr. Thomas Mgbe Inyambe for kindly reading through the draft outlines of the book and conducting accordingly a literary surgery on this work.

I cannot afford to withhold my deep admiration and profound gratitude to my father, Chief Thomas Igbodor Ibakpa Ayim and my Mother who milked my toothless gum, Mrs. Janet Ajoro Ibakpa Ayim (Nee Ushishe Agom). May God reward you a thousand fold.

Thank you Rev. Mr. Mbong Ekpo. You sat beside me as I read and re-read the manuscripts. You encouraged me. I am sincerely indebted to you for giving me a suitable choice of a title for this work.

Ishong Nkong, really I do not know howelse to thank you and your dear wife for the love you gave me. And may the Almighty God return the same love. God bless your beautiful family!

Mr. Matthew Aruku Inyambe you are an objective mind. You went through the manuscripts as Principal Nursing Officer (PNO) and scripted a very revealing "FOREWORD". Thank you! To Prof. Asindi and Dr.John O. Odok, thank you for leaving your views on this work. Finally, my sincere thanks goes to my beloved sister, Mrs. Agbo Alice Adinya who loves me all the time she hates me. This book was rightly dedicated to you. And to you reading this piece, I say thank you so much!

Ushonye Ayim
Ogoja – Nigeria
August, 2007.

PREFACE

This work, **HIV/AIDS – HOW EXPERTS SPREAD THE VIRUS** is intended to provoke the Medical Scientist to spend more quality time in the clinical laboratory more than anywhere else, to enable this generation of mankind that is so plagued by the AIDS virus receive some relief. It was not intended to provoke them or anyone else to anger or strife that this work is written. In fact, the medical community the world over would appreciate this piece of literature even the more.

HIV/AIDS enlightenment campaign is suffering from stagnation. And I have become increasingly aware of the responsibility to stir up this stagnant pool and cleanse it of its sedimentary deposits.

Teachings on the dreaded disease have fallen into much more serious error as I have discovered and now uncover to my generation without hesitation. The errors associated with AIDS (Acquired Immune Deficiency Syndrome proves a half – truth which is a whole lie. The world tends to be contented with this half-truth.

The fact is man's confidence in the reputation of the institution of medical practice has led him to accept a half-truth. Speaking candidly, investigations have proven so.

There has been no establishment of truth if the latex membrane condom is clinically recommended against the contraction of HIV/AIDS. This rather fosters its spread. Numerically, African patients constitute a large group and their governments have incurred loans from World Bank to spread butter on the HIV/AIDS expert's bread.

In this book, lie a string of questions to all those who personify the Institution of Human Medicine. These questions on HIV/AIDS will definitely help mankind to determine when a clinical clue is worth pursing or when to dismiss it as a red herring.

The idea of HIV/AIDS is terror and terrorist in itself, because the clinical history of most cases is full of error. The entire human race has been held prey, especially the sub-Saharan Africa (Black Africa) and my beloved nation – Nigeria has not been spared. Enough is enough!

Let the HIV/AIDS experts go on and answer these questions because it seems to me that the HIV/AIDS intervention agents are spreading the virus rather than a message. In fact they should tell the world why they should speak from both corners of the mouth over the dreaded disease.

In this work, questions and references have been placed side by side and most of the references were cited from the works of seasoned Medical Practitioners and the HIV/AIDS intervention literatures to further justify the fact that intervention agents speak from both corners of the mouth.

I enjoin everyone, especially those of the Medical Community to read this piece of literature with the highest standards of neutrality. As you will appreciate, there are unconquerable documented scientific proofs on the subject as reflected on the bibliography to this text.

Let my brothers and sisters in the profession of Human Medicine do more through intensive clinical research to give the world superior preventive measures and proper enlightenment about this human virus, since it has yet no cure. Existing preventive measures and enlightenment campaigns over the AIDS scourge so far, have proven inferior except the principle of sexual abstinence, which alone proves superior, and a few others like avoiding the use of unsterilized piercing objects and the transfusion of unscreened blood, that have proven reliable. But then, the superiority of 'Sexual abstinence' has been weakened by the reality of sexual promiscuity among the populace, especially the youth.

This reality, even the medical community or HIV/AIDS experts do appreciate, for this is why the latex membrane condom is clinically recommended against the contraction of the AIDS virus. Unfortunately documented infallible proofs have proven porous, therefore leaving only the virus to pass freely through its pores even where there was no condom accident or "burst". Remember, first of all that the condom is a contraceptive and is no preventive barrier against HIV or the AIDS virus.

Really, I am provoked to compassion for this generation, therefore provoking the Medical Scientist to return with spontaneous immediacy to the clinical laboratory, so that beneficiaries of the Medical Practice could reap the benefits of the profession in its fullest.

I do not; on this page(s) condemn the medical skill in anyway but only to reveal a weakness that could provoke some more strength in the muscles of the profession and consequently help in the pursuit of better remedial action against the pandemic. However, this generation should not be allowed to drown in the salty waters of shameful ignorance over some ineffective preventive measures prescribed earlier by HIV/AIDS experts. No!

Finally, there is a conviction in my spirit that many a member of the medical community may feel the same way I do, but are bound by some professional oath and confined to some laid down "professional principle" of the Medical Practice, thereby compelling the reservation of such impulses to private meditation. Once again, may I add that you read this work with highest standards of neutrality and sincerity of purpose. Have a pleasant reading!

CHAPTER ONE
HIV/AIDS: ORIGIN

Experts maintain that the real origin of HIV/AIDS is not clear. But they argue that there are two "schools of thought" concerning the origin of HIV/AIDS, one of which is traceable to African Monkeys.

It seems HIV/AIDS experts do not understand perfectly, the aetiology of the said virus, therefore do asserts still the existence of schools of thought concerning the origin of the virus. All one expect from experts is to say, 'the real origin of HIV/AIDS is not clear'. This should be their anthem and not conflict issues by recording works about 'schools of thought' regarding the origin of the virus. This is where I maintain that HIV/AIDS experts ought to have a rethink.

After telling a man you don't know the origin of a particular thing yet you begin at the same time introducing to him routes of its origin in the name of 'schools of thought'.

At that point, you conflict issues. If one may well ask, should there ever be a school of thought that could be traceable to African Monkeys? I believe no authority in HIV/AIDS should ever mention such a school of thought concerning the origin of the AIDS virus.

Here is the position of an expert on its aetiology. *'The real origin of HIV/AIDS is not clear. However, there are two schools of thought concerning the origin of HIV/AIDS. One school of thought has it that HIV/AIDS could be traceable to African Monkeys, while the other school of thought traces HIV/AIDS to laboratory accidents. This latter, while some scientists and researchers were experimenting on certain viruses, some strains that were resistant to killing by any modern chemotherapy developed'. Ofor (2003) p.5.* Experts have told us they have no knowledge about the origin of the virus and at the same time they have suggested an origin traceable to African Monkeys even where an AIDS intervention literature points up a question, *'what does each letter in HIV means? "H" stands for Human because the AIDS virus only lives in human being and not in animals, insects, water or air'. ANSWERS TO QUESTIONS ON HIV/AIDS (2003) p. 23.*

So, if the AIDS virus does not live in animals nothing about its origin should be traced to any monkey, whether African, American, or whatsoever. Monkeys are animals, isn't it? I suggest conflicting assertions like these are no good, therefore proper Clinical research should be reinforced to save my generation from conflicting information over the origin of the said virus.

CHAPTER TWO

THE LATEX MEMBRANE CONDOM

Should HIV/AIDS experts recommend consistent and correct use of the condom, where the latex material has natural tiny pores (holes)? The condom is advertised all over the world as the recommended preventive barrier against AIDS virus. The world has been deceived into believing that provided there is no condom accident or rupture during the use of the latex material, you could well have safe sex.

This is one aspect the HIV/AIDS intervention agents spread the virus rather than a message. How could you teach the world that using condom during sex prevents one from contracting the virus even where the condom never got burst? This also, is a frank lie. Some rate the safety of condom in percentages. The condom is made from latex. And since it is a membrane, it has natural tiny pores or holes. Therefore it does not command safety, not even to a single digit.

4

If it does, it is zero percent. Correct and consistent use of the latex membrane condom gives nobody whether male or female any protection against the virus. *'It is easier for the AIDS virus than for sperm to pass through the condom pores (natural tiny holes in the latex material) because the virus is 400 times smaller than a sperm'. McSweeney (2001) p. 64.*

From this, one comes to a sure knowledge of how porous the condom is even where there was no condom accident.

There is no condom in the whole world that keeps one from contracting the AIDS virus. Every condom is made from a latex material. *"A male condom is a sheath of thin plastic or polyurethane rubber or natural membrane". ANSWERS TO QUESTIONS ON HIV/AIDS (2003) p.51.* This is the natural membrane, Dr. McSweeney revealed earlier in this chapter as carrying tiny pores wide enough for the virus to pass through from an infected person to a non-infected person even where there was no condom accidents.

Why do the HIV/AIDS experts recommend the latex condom against the contraction of HIV/AIDS where it has natural tiny pores or holes? HIV/AIDS experts have been spreading the virus through recommendation of the latex condom. The world should be told the truth. AIDS intervention agents should begin to teach people the truth about HIV and the condom, and how its protective ability is zero.

Even where you use a condom without bursting, it were as good as not using one, if it were intended to prevent you from contracting or transmitting the virus. So many AIDS intervention literature advertise consistent and correct use of the condom against the contraction of the AIDS virus. And even where it is imprinted "electronically tested" on condom wrappers, never feel it has any protective ability against the virus. In another occasion, if the manufacturers of the condom imprint 'sealed' on packs carrying the latex material, do not get distracted. The HIV/AIDS experts should apologize to the world for recommending a contraceptive (condom) as preventive barrier against the contraction of the AIDS virus.

The HIV/AIDS experts have even developed a slogan known as the "ABC" of HIV/AIDS. The 'C' meaning consistent and correct use of condom, the "C" aspect of the intervention approach to safer sex regarding the AIDS virus is defeated since one can contract the virus even where the condom is correctly and consistently used during sex (penis to vagina). So, why has the AIDS expert deceived the world into accepting the latex condom as a recommended preventive barrier against HIV/AIDS?

CHAPTER THREE

SALIVA AND THE MUCOUS MEMBRANE

If there is a little quantity of the virus in saliva, is this little quantity incapable of permeating through the mucous membrane (smooth lining or invisible crack) of the mouth since the virus can pass through this lining even if it were not injured?

In a HIV/AIDS intervention literature, we understand it will require a gallon of infected saliva to infect a healthy person through a kiss involving saliva exchange. *'The quantity of HIV in saliva and urine is so tiny that it cannot cause infection. You would need to drink a gallon of saliva (half a bucket-full) to be infected'.* *McSweeney (2001) p.27.* The question is, are we concerned here with the quantity of saliva exchanged or with the quantity of the virus? Because, reliably a single virus is capable of infecting someone since it replicates or reproduces in the human body.

So, since a single virus can infect a healthy person and replicate as it enters and resides in his or her body, why do we talk about the quantity of saliva. We should talk about the quantity of the virus and a single virus can infect. So if one may well ask: what then is the HIV/AIDS expert teaching?

In a kiss involving saliva exchange, an AIDS interventionist supports the fact that the virus can pass through the mucous membrane of the mouth, that smooth lining that flakes off when very hot tea is sipped (it is waterproof-like) *"The mucous membrane is the inner lining of the vagina, penis, anus or mouth and the virus can pass through this lining even if it is not injured"* McSweeney (2001) p. 27. So, if a single virus can pas through the mucous membrane of the mouth and infect a healthy person, what is the AIDS intervention expert preaching?

It means a condom for the mouth is required, since through the mucous membrane a virus could pass through and infect someone during a kiss as it is usually described.

If a "condom for the tongue" is not developed for public consumption, it means HIV/AIDS experts are spreading the virus rather than a message, since it no longer requires a large quantity of saliva but just a single virus to infect someone through the smooth lining (mucous membrane) of the mouth.

In a Reproductive Health and HIV/AIDS prevention booklet, we learn one could get infected through invisible crack or bleeding gum during wet or deep kissing. *"Yes, Scientists think that the AIDS virus can pass through open sores or cut or invisible crack or bleeding gum or tongue in the mouth of infected person to another during wet or deep kissing in which a lot of amount of saliva is exchanged.' ANSWERS TO QUESTIONS ON HIV/AIDS (2003) p. 36.*

Now, what else is the invisible crack? The only invisible crack in the mouth even where there is no bleeding gum is the mucous membrane with its invisible pores.

If the AIDS expert knows that the virus can be transmitted through invisible cracks (mucous membrane) in the mouth, why then do they not teach so and develop, as I said earlier a 'condom for the mouth or tongue' rather than spread the virus as they have done and still do?

In reacting to the question, 'Is there a link between oral sex and HIV/AIDS? One Health Education booklet argues thus, *'It is a low risk route of transmission; the HIV is killed by the acid in the stomach'. Sexwise Guide (2000) p.30.* Now, considering the fact that a single virus can pass through the smooth lining of the mouth, even when it is not injured, and infect someone, how else in a situation of sucking either infected fluids of the male semen or vaginal secretions in the female? It stands to reckon, therefore, that the HIV/AIDS expert is spreading the virus rather than a message. You argue that oral sex is a low risk route? No wonder, this teaching has kept lesbianism on the rise, such an unnatural sexual attraction.

And I make bold to say here that, one of the fastest routes of HIV/AIDS transmission is through oral sex, unless there is an 'Oral sex condom'. And do not forget, the condom lack protective ability in all its forms.

In the Sexwise Guide, we learn the acid in the stomach kills the HIV. The HIV/AIDS experts tell us that the virus can survive anywhere except in insects or animals, why then is the virus killed in the stomach which is a part of the human body? By the way, why can't they (The Medical Scientists) develop a synthetic acid of the type found in the stomach that kills the virus as permanent vaccine against the scourge? These HIV/AIDS intervention agents do very much seem to spread the virus rather than a message. And what is the meaning of all these?

CHAPTER FOUR
CONDOM FOR THE TONGUE

Since, a little quantity of the virus is contained in saliva, don't you think the introduction of 'condom for the tongue' will do?

Yes, just as I had mentioned earlier somewhere in this volume, about 'condom for the mouth', which may perhaps help in preventing a healthy person from catching the dreaded disease, I still maintain so here.

The truth is, since we reliably understand that a little quantity of the virus is contained in saliva and understand still that even a single virus can infect someone through the smooth lining of the mouth, I think what we should ask for from the AIDS expert is the introduction of the condom that could be worn in the mouth or over the tongue as a preventive barrier against the virus. Meanwhile, there is just no way a kiss involving the exchange of saliva from an infected person will not be passed on to a healthy partner without a "mouth condom".

Let me state here once again that, the transmission of the virus is made possible through wet kiss, because the mucous membrane of the mouth is porous even without any injury. *"The mucous membrane is the inner lining of the vagina, penis, anus or mouth and the virus can pass through this lining even if it is not injured'. McSweeney (2001) p. 27.*

Really, the introduction of a condom for the mouth or tongue will be as good as rubbish because even the condom worn over the penis or inserted into the vagina (female condom) is made of a latex material. The latex material has natural tiny pores or holes which leaves it incapable of preventing the virus except one is deceived to accept a lie as I see millions of people do, believing this half – truth from AIDS literature/experts, that the latex condom prevent healthy people from contracting the dreaded virus.

Developing a latex membrane condom against the mucous membrane in the mouth is like putting a lick against a lick, because the latex material is naturally porous or semi-permeable, just like the mucous membrane itself. It is indeed a pathetic situation. The situation calls for caution, very serious caution. Meanwhile, let the HIV/AIDS intervention experts once again apologize to the world for not recommending earlier the introduction of a latex membrane condom for the tongue or mouth. Really, I want the HIV/AIDS expert to tell this generation of mankind the truth about this dreaded disease, because it seems to me that there is some skeleton left in the cupboards concerning the AIDS scourge.

What is the meaning of all these? People, so many people must have gotten infected through infected saliva and the mucous membrane in the mouth where the virus could pass through even when there is no bleeding gum or sore in the mouth.

This book is a timely intervention against the many contestable postulations from AIDS experts. People should know the various routes they could contract the dreaded virus than be allowed to go the way of these disputable claims. They are in deed spreading the virus rapidly. There is no message if the AIDS experts insist that only a large quantity of infected saliva can infect someone without considering the fact that the presence of a single virus in saliva be it a whole jar, leaves that quantity infected.

If you really do not want to catch the dreaded disease, stay away not only from unprotected casual sex as they preach but also from unprotected "wet kiss', because even if a latex condom were developed for the tongue or mouth, it will still carry natural tiny holes which are invisible to the eyes. Meanwhile, is it a large volume of infected saliva that is required to infect someone or large volume of the virus? By the way, can't a single virus infect one?

Finally, I suggest still, the development of a condom against unprotected oral sex. A pamphlet on AIDS/STD control supports my suggestion of a "condom of the mouth". *"There is also a risk of contracting HIV in <u>Unprotected oral sex,</u> or anal intercourse if one of the partners is infected".* *AIDS: What Teachers Should Know. p.4.* Enough is enough!

CHAPTER FIVE

THE DISCORDANT COUPLE

Why is it that a man tested positive and died from AIDS while his spouse (wife) tested negative even after the "window period"?

There has been so many oral testimonies over the fact that a particular partner is diagnosed for HIV/AIDS and found positive, even dies from it, whereas his spouse remains healthy even after the window period or incubation period. Discordant Couples, right? May be! Well, no problem!

Confirmations on this issue have reached us even from the lips of Medical Practitioners themselves. HIV/AIDS experts in their campaigns have always contended that an easier route of transmission of the dreaded virus is sex. Some intervention agents argue that if one goes through a woman after she is wet, he is not likely to contract the virus. Reliably the virus, we understand, is found in body fluids such as vaginal secretions, semen, blood, etc. *"During sex, the AIDS virus which are too tiny to be seen with the naked eyes except through special magnifying glass known as microscope can pass from one infected partner to another in any of the sex fluids (semen or vaginal fluids) even if you have sex only once or for a short or long time".* ANSWERS TO QUESTIONS ON HIV/AIDS (2003) p.32. So, how come one does not contract the virus during sex only if the infected woman is allowed to "wet", whereas the very fluid that made her wet is resident in the vaginal environment. Someone may argue that after a woman is wet, one could get her douche or wipe out the secretions before actual penetration, thereby keeping one safe from the virus. *"This is not true, although some women believe that <u>douching</u> or <u>washing</u> the vagina immediately after sex will prevent getting infected. But unfortunately this act increases the chances of getting other infections to the reproductive organs and also the <u>AIDS Virus</u> even if <u>it is done before or after sex?</u> ANSWERS TO QUESTIONS ON HIV/AIDS (2003) p.28.*

May I reiterate here once again that those who believe getting an infected female partner "wet" before penetration as a means to avoiding being infected are dead wrong. Even if they feel it is only where there are sores in the vaginal environment of an infected female partner that transmission could occur, is indeed a disputable claim. This is so because a thin lining (mucous membrane) covers the vaginal environment, which is wide enough for the virus to pass through.

People ought to be told the simple truth about the dreaded virus without hesitations. Even without sores in the vagina or penis, sex with an infected partner could still leave one prone to the Virus due to the presence of the mucous membrane.

Some individuals have what Genetic Science call the Chemokine Receptor (RC), attached to their blood cells (CD4), while some other individuals do not. This receptor is what attracts the AIDS virus to course infection in one partner while the other stays healthy.

Earlier on, the issue of the 'Discordant Couple' aroused a lot of anxiety in the mind of the average, and even the scientist. Some has submitted that the issue was a medical mystery. Today, some argue that, for the scientist to trance the issue to the 'Chemokine Receptor' simply unfolds the mystery. They run the conclusion that the medical mystery of the 'Discordant Couple simply stands demystified.

Really, the situation where a man test positive while his spouse test negative, proves that the spouse's CD4 cells do not have the receptor to attract the virus. So, in the event, no virus is permitted in. And this goes a long way to teach the rest of mankind that the issue of the 'Discordant Couple' is no longer medical mystery, it is of course, medical history.

CHAPTER SIX
LESBIANISM AND HIV

The sex mode employed by lesbians is predominantly the "mouth-in-vagina" sex route. Where is the possibility that the ratio of contracting the virus through this sex route is so small as compared to homosexuality (Penis-in-anus) and the vaginal sex (Penis-in-vagina)?

AIDS experts maintain that the chance of lesbians contracting the virus is little or zero as compared to homosexuals. This is where I wish to differ. For instance, one AIDS interventionist contended that, *"Lesbianism is a low risk route of transmission of the virus". McSweeney (1995) p. 40.* And in another occasion we are told oral sex is risky *"Included here are those who practice "Oral sex" which a small number of people use instead of vaginal intercourse in order to avoid pregnancy, (Homosexuals sometimes use it)* <u>*There are a risk of getting AIDS in this way because semen contacts the mucous membrane of the mouth'*</u>*. McSweeney (2001) p. 42.* So, it is not entirely plausible for HIV/AIDS experts to say that lesbianism (mouth-in- vagina sex route) is a low risk route.

AIDS Interventionist maintains that infected vaginal fluids contain the virus and the mouth has a mucous membrane wherein the virus could pass freely to infect a healthy person, simply shows that they (the HIV/AIDS experts) are rather spreading the virus. Where is the possibility that sucking out infected vaginal secretions or fluids will not get every lesbian infected when in contact with virus-ladened vaginal fluids?

One other AIDS intervention literature argues thus, ... *"The HIV is killed by the acid in the stomach". Sexwise Guide (2000) p. 30.* I wish to differ. Even if there were acids in the stomach that killed the virus, I'll like to let us know that even before the infected vaginal fluids leave the mouth for the stomach you must have gotten infected through the mucous membrane of the mouth which allows the free passage of the virus through the walls of the mouth even if it were not injured. Oral sex, which is the lesbian sexual route, spreads the virus even faster. *"There is also a risk of contracting HIV in unprotected oral sex, or anal intercourse if one of the partner is infected".* AIDS: *What Teachers Should Know* p.4.

CHAPTER SEVEN
RELIGION AND DISCRIMINATION

Religion is one of the factors that affect a person's sexuality. Why are couples being denied getting wedded even where a healthy partner is ready and willing to marry the infected one? Should we treat or regard HIV positive status as a disability? Is there any prohibition against the Disable in any Religious cycle? And do you suggest Mercy-killing for those living with the virus? So, why all these?

This is a question of discrimination. We all know love is strong. For instance, in a "White wedding", its content is the exchange of vows or to swear to an oath, the oath of marriage. The marriage oath consists of words the couple repeats after the clergy one after another. The oath is usually nicknamed "For better; for worse". From this, it shows that marriage ought not be interrupted if the theme of the vow is "For better; for worse', even if one partner is infected.

I remember the marriage oath is recited thus, ... 'take this ring as a sign of my love and fidelity, for better; for worse, for richer; for poorer, <u>in sickness and in health,</u> till death do us part.' Yes, this is the vow usually exchanged before the clergy at the altar. So, if one partner is sick and the healthy one is still ready and willing to marry that partner, why should they be discriminated upon? Are we deviating from the vow, "…In sickness and in health'?

The Religious Community is exerting a form of discrimination over the HIV/AIDS subject. Heal the couple and wed them. Don't discriminate against them or expel them from the congregation if the healthy partner insists on marrying his or her sick partner due to the "Strength of love". After all marrying isn't necessarily all about sex. And even if it were so, healthy babies could still be born to discordant couples through the PMTCT (Prevention of Mother To Child Transmission) programme.

Someone once told me that it is only when a couple is already wedded before contracting the virus that the spouse is bound by the marriage vow. Was he right? Certainly not! I am very much convinced that the clergy have acted many times this way without knowing that they contradict themselves and the vows of marriage which is usually recited after them by couples at the Altar. This form of religious discrimination is quite unfair.

CHAPTER EIGHT

AIDS: CONGENITAL OR NON-CONGENITAL?

AIDS simply means, "Acquired Immune Deficiency Syndrome". 'Acquired' in the sense that the condition is not congenital. But then, we learn the unborn child gets infected with the virus.

The Oxford Advanced Dictionary of current English defines the word "Congenital" as *'of diseases present, belonging to one, from or before birth'*. Also, the New Lexicon Webster's Dictionary of the English Language (Deluxe Encyclopedic Edition, 1991) defines it as, *'Present at birth, acquired during fetal development and not hereditary'*. So, having undertaken semantic investigations into the word 'Congenital', we could now begin to compare notes. Where a disease(s) is present with one from birth or even before birth, such disease is described as congenital.

HIV/AIDS experts maintain that the disease or condition (AIDS) as they put is not congenital. *"AIDS simply means, Acquired Immune*

Deficiency Syndrome. Acquired in the sense that the condition is not congenital, but comes as a result of a change or changes in the individual's Immune status that tends to allow easy attack by some other disease – causing agents.' Ofor *(2003) p.1.* From our earlier definition of AIDS, it becomes practical that nobody can contract the virus in the womb to cause it suffer a diseased condition known as AIDS.

But, think I see something quite different from the impression built earlier on. This is where I become informed to put this work together for public consumption *"various means of the spread of HIV/AIDS are known, namely:*

1. *Sexual Transmission*
2. *Blood Transmission*
3. *Parental Transmission*

Ofor (2003) p.6. From this assertion, there seems to be a contradiction. Now we learn "parental transmission" is one of the various means of the spread. Why these conflicting assertions from the works of experts? What really are they telling the world? Now see this,

"Women with the virus can pass it on to their baby during pregnancy, at birth or via their breast milk'. Sexwise guide (2000) p.28. This assertion proves that it can be acquired from the womb, leaving the doctrine of HIV/AIDS "Not being congenital" false and very confusing.

Further revelations assert that HIV/AIDS can be transmitted, *'...from infected mother to her unborn child during pregnancy or during breastfeeding'. Answers to Questions on HIV/AIDS (2003) p. 32.* Further on, on page 48 of the same book, momentum is added to all indications, the HIV/AIDS experts are saying that the disease is congenital and non-congenital.

We have been told earlier as I did point out, the HIV/AIDS expert had said in some other occasions that the blood of the unborn baby and that of the placental barrier cannot allow the transmission of the virus from an infected mother to her unborn baby. This is what we are told and this is where I believe HIV/AIDS experts speak from both corners of the mouth.

Now, to the same experts, when asked how is it possible for a pregnant woman who has the AIDS virus to pass on the virus to her unborn baby, had this to say, *'This can happen during pregnancy when the virus passes through the placenta to the baby before/during child birth and after delivery during breast feeding? Answers to Questions on HIV/AIDS (2003) p. 48.*

The assertion that an unborn baby could get infected while in the uterus. So, this is where I ask that HIV/AIDS experts return with spontaneous immediacy to the clinical laboratory for superior findings rather than humiliate the profession themselves by asserting that HIV/AIDS is 'Not congenital', whereas there have been obvious traces of the unborn baby contracting the virus even while in the confines of the uterus of an infected mother. Really, the profession of human medicine still needs to do more over the AIDS scourge in its enlightenment campaigns throughout the world.

Further confirmations that HIV/AIDS is rather 'congenital' states, 'In passive transmission', a mother can pass AIDS to her child before or during birth'. The clinical history of the disease should be reviewed.

CHAPTER NINE
HIV AND BREASTMILK

Is it possible that a single child who ever sucked gallons of infected breast milk live without the virus? AIDS interventionists seem to argue that it's possible. *"But unfortunately if the mother is HIV positive there is now clear evidence to show that in 20 percent of such mothers HIV is passed on to their babies in the breast milk, causing one out of five babies to develop AIDS. In 80 percent of infected mothers the amount of HIV in their breast milk is so small that it does no harm to their babies and four out of five of such babies do very well on breastfeeding. In the remaining one out of five the amount of HIV in the milk is much increased causing their babies to be infected with the virus and later to die of AIDS'. McSweeney (2001) p. 50.*

Is it not these AIDS experts that told us earlier that the adult will require a gallon of saliva to contract HIV/AIDS in a kiss involving saliva exchange? Now, here is a baby that sucks gallons of infected breast milk from an infected mother and the same expert is telling us that not all babies who suck infected breast milk can get the virus. By the way, have HIV/AIDS experts not exposed to us how an acid in the stomach kills the HIV? ..."*The HIV is killed by the acid in the stomach*". *Sexwise Guide p. 30*. And if one may well ask, is it the absence of the said acid that causes some kids to get infected? The notion that some kid do not get infected from infected breast milk while others do is quite funny, very funny. Meanwhile, every kid has in its mouth the mucous membrane that can allow even a single virus in, from infected breast milk, since it is in itself semi-permeable. Really, the expert needs to do more work.

CHAPTER TEN
HIV AND LEUKEMIA

Is HIV/AIDS another variety of the Leukamia or is it an old wine in new skins?

Yes, Is HIV/AIDS another variety or a new form of Leukemia that was first reported widely in 1981 in the United States of America (U.S.A)? Is it a new form of Leukamia with the peculiar character of being contracted sexually among other things?

Leading Medical Specialists around the world have made it clear to me that Leukamia is false cancer of the white blood cells. *'Because Leukamia involves blood cells circulating through the body rather than a fixed mass of tissue. Leukamia is sometimes not considered a true cancer. However, Leukamia cells, when studied under the microscope and in cell cultures, behave like cancer cells found in tumours'. Medical and Health Encyclopedia (1997) p.317.* This assertion declares that Leukamia behaves like cancer cells found in tumours just as *'AIDS generated cancer' medical and health Encyclopedia (1997) p.320.*

Is HIV/AIDS a new variety of the various forms of Leukamia? Why is it that common symptoms to all Leukamia are associated with AIDS? *'Common symptoms to all Leukamia include fever, weight loss, fatigue, bone pain, anemia as is expressed in paleness, and an enlarged spleen or masses of Leukamia cells'. Medical and Health Encyclopedia (1997) p.317.* Assertions like these, brings one to the brink of understanding that AIDS is a new form of Leukamia that has the peculiar character of being transmitted sexually.

Is it true that when the white blood cells are attacked it leaves the patient unprotected from germ invasion that could result in opportunistic diseases? For example the lungs, the intestine, the throat, the skin and the gums are the first lines of defense against infections – This is usual with Leukamia.

It may interest you to know that the HIV/AIDS expert had concealed the original name given to the virus from the lay public as it is confined to only the medical literature. The original name of the AIDS virus however, makes it clear, even clearer that HIV/AIDS is a new variety of the Leukamia discovered in the USA. *'First reported widely in 1981, AIDS has become a priority of the U.S. Public Health Service. Researchers have isolated a virus, the human immunodeficiency virus (HIV), that they believe causes AIDS. This virus was initially called the human T – cell Leukamia virus III (HTLV-III)'. Medical and Health Encyclopedia (1997) p. 315.* Yes, it's the Luekamia virus type 3.

This makes it quite clear that HIV/AIDS is a new form of the Leukamia that could be sexually transmitted. By the way, why is its original name concealed from the lay public? Does this change in names not prove that the original name of the virus, Human T – cell Leukamia virus III (HTLV-III) is the old wine while its new name, 'the Human Immune Deficiency virus ' (HIV) is the new wine skin?

APPENDIX ONE
GOD AND THE AIDS VIRUS:
WHO IS MORE FEARED?

The big question is, between God and the AIDS Virus: who is more feared? The latter seems to be more feared, I guess.

Today, Religion is being involved in combating the AIDS scourge through the faith based 'approach as it is called. Some see this approach as fanatical. For instance, the 'faith based' approach condemns the use of the latex condom because to them it encourages sexual recklessness among the youth and the entire adult populace. This approach is only meant for convicts in the religion. Every true convert in a religion where promiscuity is frowned at, would not need anyone anywhere to teach him or her that the use of condom is no good but abstinence.

Today, the latex membrane condom has been scientifically adjudged porous i.e. to have natural tiny holes such that the virus could pass through even where they were no condom

38

accidents. Now, those within the secular who believed in the use of the condom but condemned the 'faith base' approach are now preaching the 'abstinence' or 'Zip – up' approach, because they've realized the condom now has tiny natural holes wide enough for the virus to pass through.

Really, the 'faith based' approach today is a superior prescription within secular circles since the condom is adjudged scientifically porous even with the certainty of sexual promiscuity in our society. Today, I see most people choosing the 'Zip – up' approach as the best option in guarding against the contraction of the virus. At that level, it shows that most people have chosen abstinence not for fear of God but for fear of HIV or the AIDS virus; not for love of God but for love of self, not because they realized according to Apostle Paul that their body is 'temple of the Holy Spirit, but because they don't want their body to be temple of HIV/AIDS. So the question still lingers on, God and the AIDS virus: Who is more feared?

If salvation were based on sexual morality, then for all whom because of fear of HIV/AIDS have chosen sexual morality should be congratulated as having wrought salvation. Could you imagine that? So, at the end or at judgment, when everyone must have led a sexually moral life for fear of AIDS and not for fear of God, how would it seem? I think it would become a matter of 'motive' or intention. What is the motive behind this sexually moral life you led? Was it for God or for self? Was it because the Scientist declared that the latex membrane condom has natural tiny holes? Where do you stand? It is quite true that, because of the porous nature of the latex condom, the only option open to the secular world is 'abstinence'. God searches the heart. He knows all of our intentions. He knows where you choose sexual abstinence for fear of AIDS and where you choose same for fear of Him. But then, God is not judging us into heaven for acts of moral righteousness.

Let me warn all those who say, God allowed HIV/AIDS to compel humans into leading sexually moral lives. Such persons are equating the love of God for fear of HIV. That is to say, if you fear HIV, it means you fear God. It therefore means God is reduced to a virus. So long as you fear this virus you fear God – this is unacceptable! My God is not a 'virus'. Fear of this virus comes nowhere near the fear of God.

May I make this point clear, that all those who jump up all at a sudden, because they were heart broken by a former boyfriend or girlfriend, thereby becoming emotionally frustrated, because of this choose to lead sexually moral lives in the name of old things have passed away' or being 'Born again', should reexamine themselves to know if they where driven by a true love of God. I am in sympathy with a lot of preachers who mistake people like these for true faithful. God won't make such mistakes. He knows the heart. Some come especially ladies, when they are out of circulation (on the bad side of 30 or 40 years of age), in quest for husbands. Certainly, the sheep will be separated from the wolves.

May I warn all preachers in religious circles that they should not feel or believe that all who adopt the doctrine of 'sexual abstinence' are doing so for fear of God but for fear of AIDS. It is the motive and not the action that differentiates a hypocrite from a faithful. The pointed question is: Have you adopted sexual abstinence for fear of AIDS or for fear of God? Think about it!

APPENDIX TWO
'ZIP – UP'
(THOU SHALL NOT FORNICATE)

IS A
SCIENTIFIC PRESCRIPTION
AGAINST
HIV/AIDS

Yes, all those who have been cajoled in religious circles as fanatics, because of the prescription of sexual abstinence for the youth can now heave a sigh of relief, since Science and the Health Scientist now prescribes "Zip – up" (sexual abstinence) as a veritable tool against the spread of the dreaded virus.

Before now, only moral preachers, sorry, preachers of simple morality were advocates of sexual abstinence but today the story has changed due to the semi-permeability of the latex membrane condom in the face of the HIV. Today, I perceive that the fight against the scourge and the failure of the condom in combating the spread of the virus has entrenched in Science, the Scientist and the Scientific world what I describe as 'scientific sexual morality'. The Scientist has given the campaign for sexual abstinence the baptismal name of "Zip – up" to connote sexual abstinence. Is it a new register in the Science parlance? Well, whatever it is, the truth is that, be it the scientific "Zip – up" or the simple religious "sexual abstinence" both point at curbing the spread of the virus and not necessarily doing God service. So many people especially the youth will no longer fornicate if the preaching of Pastors could not keep them away from fornication rather, the sermon of the Health Scientist titled "Zip – up" will do. And as such all those who refute the preaching of the Pastor earlier but only to accept a Scientific sermon later, have only proven that sexual abstinence is mere "self righteousness", and nothing more. Because were it for the "Love of God" the preacher's preaching could have been respected and imbibed upon but it is for the "Love of self". So the Zip – up sermon of the Scientist is most welcome.

Has HIV/AIDS become a parallel god with it preachers (The Health Scientists) and with a sermon (Zip – up)? The religious preacher is so busy preaching sexual abstinence, which few adhere to and if brought under microscope viruses of hypocrisy could be found within them. This indeed is funny.

It appears that while the Heath Scientist is preaching the Zip – up sermon to convert souls to a healthy living and consequently keep them away from the HIV, preachers of simple morality (Sexual abstinence) have rather met the misfortune of spreading the 'virus of hypocrisy' in their converts who most of the time are victims of sex-related hypocrisy. It is either these sex related hypocrites resort to masturbation (self sex), homosexuality or lesbianism at least to justify the presence of the hypocrisy virus. At the end of the day, the question is, what service has the sex – related hypocrite done God?

The big question is, now that the Scientist is preaching sexual discipline or abstinence (Zip-up) to include homosexuality, lesbianism or whatsoever, should we now jump up and shout, "Hallelujah, I am born again"?

Remember, where the scientist says, 'Zip-up', it means one can't even masturbate because where the Zip is up, how do you reach out to masturbate? Though masturbation has not been scientifically proven as a route in the spread of the virus. May be in some future. Right now Zip-up is Zip-up!

The prescriptive truth is, today due to the prevalence of HIV/AIDS; the health Scientist is more of a moral preacher than the ordinary pulpit preacher. Think about it!

APPENDIX THREE
A SURPRISE CURE
FOR
HIV/AIDS:
THE HEALTHY MOSQUITO

Can one indeed learn from science? Gottfried Benn gives an answer, *'Although Science taken as a whole is a nuisance, one can still learn from it'.* In this postulation, we shall discover that the mosquito taken, as a whole can be dangerous, in spite of this, one can learn from the Medicinal properties of this dangerous but important insect. We could all recall that millions of lives are being lost to the malaria parasite – the mosquito's venom.

Really, it was not the intention of the mosquito to sweep mankind into untimely graves but this misfortune had occurred because the mosquito was itself sick. It is the unhealthy mosquito that transmits malaria parasites. If the mosquito were to stay healthy all its life there would be no such thing as malaria or someone dying of it. No! It seems to me that the mosquito decided that so long as it stays unhealthy, mankind would remain unhealthy too. And any moment it enjoys good health man too should enjoy same.

I believe, but not for the poor health of the mosquito, the insect would have been the most friendly of all insects known to man. The mosquito that is the most frequent blood-sucking insect decided not to infect man with HIV/AIDS – The world's most dreaded disease. Dreaded because it has yet no cure. Little did the mosquito know, that this deep love for man that led it to spread malaria and not HIV would turn around to make it the lasting cure for the scourge. Today the mosquito through this postulation has presented itself as the long awaited and indeed the 'almighty cure' for HIV/AIDS.

How did it all happen? It is, of course the big question lurking within your mind so far you've read this piece of discovery.

Yes! While I was growing up, each time I looked at HIV/AIDS intervention literature, I saw things differently and still do. Then I said to myself, *'In order to play a useful role nothing less than a scientific revolution is required'*. Hamelink,(1981)p.34. And 'Scientific Revolution' requires that *'an old paradigm must be left behind, scienticism must be betrayed, new ideas must be developed and an intellectual space must be made for a new conceptual system that suspends, or clashes with the most plausible theoretical principles, and introduces perceptions that cannot form part of the existing perceptual world'*. Hamelink, (1981) p.35. 'Revolution' within the domain of science requires frankly that new ideas be developed. And you know what! *'When new ideas are developed, value-consensus is threatened and the legitimacy of the dominant order is undermined'*. Hamelink, (1981) p.34 This is exactly what you'll find at the end of the day in this piece.

'New ideas' as you may appreciate *'express changing values and if incorporated in the concrete action of social groups, they force dominant values and their representative institutions towards either blatant violence or change'*. Hamelink, (1981) p.34. This is the blatant change I am uncovering in this postulation. We should no longer watch *'Science exhibit reluctance rather than full partnership with man in the adventure of living'*. Hamelink,(1981)p.35 Something needs to be done. And quite urgently.

Distinguished reader, to give research supreme position, I present to you a 'cure' for AIDS in the healthy Mosquito. I know it's quite unreasonable to expect any layman to believe that mosquitoes do not spread AIDS. And I know it will amaze not just you but the rest of the world that the same mosquito (Though the healthy one) is the 'Cure' for AIDS. HIV/AIDS interventionists have declared that the mosquito does not inject blood; therefore it does not cause AIDS. One of such interventionist stated thus: *'the mosquito causes malaria, but it does not cause AIDS because;*

1.   *It sucks out blood. It injects only saliva.*

2.   *The virus, HIV, cannot develop in the mosquito's saliva.*

3.   *The virus cannot develop in the acid of the mosquito's stomach. It can only multiply in human cells'. McSweeney (2001) p. 56.*

Now, this is how the mosquito became cure for HIV/AIDS. Given that the saliva glands' and the 'acid' in the mosquito's stomach are a ready arsenal against the dreaded virus, why then do we look elsewhere for a cure against AIDS.

Why should the Medical Scientist hesitate in breeding healthy mosquitoes under quarantine conditions to enable the extraction of the mosquito's saliva and its 'acid' in the stomach? These two substances within the body of the healthy mosquito have proven toxic scientifically to the dreaded virus. These substances could equally be developed synthetically, having known the chemical constituents that make them. Doctors can now prescribe a healthy dose for patients to enable quick and early recovery upon this discovery.

Sometime, I browsed the Internet and saw how houseflies were bred under quarantine conditions because of its medical properties so discovered to treat a particular ailment. Is the honeybee not bred under controlled conditions to produce honey for human and industrial consumption? The same should be done within the shortest time possible on the healthy mosquito to extract its almighty saliva in the mouth and its almighty acid in the stomach, since the HIV cannot develop in the presence of these substances in the mosquito.

Are you still looking elsewhere for a cure to HIV/AIDS? The cure is indeed in the 'Healthy Mosquito'? The Medical Scientist should return with spontaneous immediacy into the clinical laboratory to enable the production of this drug found within the tiny frame of the healthy mosquito. I suggest that after this is done, all nations on earth should pick up a date and declare it work-free to the 'Glory of God and for the betterment of mankind' for this discovery. The date is simply a celebration of the 'death' of HIV. A cure, at last is found, may be!

APPENDIX FOUR
TEST KITS:
HOW THEY SPREAD THE VIRUS

Throughout the World, the Campaign against the spread of HIV/AIDS include sexual abstinence (Zip-up), proper application of the latex contraceptive (Condom), use of sterilized piercing implements, use of screened blood, and so on, and so forth. But nothing, nothing was ever mentioned about the proper use of test kits.

The test kits in many cases spread the virus like sexual recklessness. Many people living with the virus go about spreading the virus even after a test, because the test kits gave them the go ahead. Yes, they tested negative, whereas they are positive. Does a situation as this not show that the HIV is spread through the test kits?

There are no machines to tell when humidity conditions alter the effectiveness of test kits like Statpak and Determine which are the common test kits available in Africa. May be the

Machines, the kind I suggest are available else where. But in Africa I have not heard of one yet.

I understand there are some test kits that are cold chain, i.e. must be kept in a refrigerator or under similar situation. One of such test kits, I understand is cappillus. I also understand that a test with cappillus is often referred to as the tie – breaker test.

Because where the "Determine" and "Statpak" test kits seem to differ in results, Cappillus which is cold chain does the tie-breaking. It tells us which actually got the reading right.

It is unfortunate that most African Nations do battle power outage. Yes, epileptic electric supply. This is a very serious threat to the preservation of certain caliber of test kits. This is entirely Africa's short coming. Africa needs to address her energy problem, since it goes a long way to impact on the health of her peoples positively. And I make this appeal to our sister continents to please come to Africa's aid and in the energy sector.

I also wish to challenge Medical Experts in Africa. The first purpose of the Practice I perceive is the Patient's health condition and how it could be turned around.

Many times I see Medical Practitioners go on Industrial Action in pursuit of pay rise. Yes, "the Labourer deserves his wages". But why can't this Labourer protest against poor working tools? Tools that could inflict terrible blisters on the very hands with which the job was done.

I want to see Medical Practitioners in Africa embark on Industrial Action, this time in protest of poor working tools. I see too many blisters on their hands. And it gives me quite a great worry.

A Doctor has a case in his hands which he can treat, but because of poor energy supply, he looses that case to the Mortician!

Now, I understand there are sophisticated HIV Test machines that rarely exist in Africa and some parts of the world. For example, the Viral-load Test machine which detects the level of white blood cells (CD4) in the body, which will determine if a positive case could be placed on ART (Anti-retroviral Therapy).

The PCR (Polymerase Chain Reaction) or DNA (Antigen) Test Machine is another one. This one detects the presence of the virus even at "window period", i.e. the period where HIV antibodies have not yet developed in the blood.

You can imagine the damage associated with the absence of these machines.

Don't you realize that if the antigen test is conducted, and the virus detected in time, at that level it saves the spread? So at that, the absence of these test machines will rather foster the spread of the virus. And this is where I think Test kits in their different categories whether unavailable or poorly preserved could go a long way to aggravate the scourge. Something has to be done, and quite on time!

I use this medium to also appeal to African Governments to see the economic damage a poor energy sector can leave on an Economy. "A healthy nation", they say, "is a wealthy nation". If our best brains get sick and die, what happens to our economy? African leaders do something to step up our energy sector now!

APPENDIX FIVE

POSTSCRIPT I

New facts, ideas, diagnostic capabilities, drug policies and prevention keeps emerging so often on Human Immunodeficiency Virus (HIV) and Acquired Immune Deficiency Syndrome (AIDS). Frequently, the information and policy changes regarding the infection can become quite controversial. These controversies and seeming confusion therefore provide a fertile ground for the skeptics to challenge the medical scientist. In this book whose contents are disposed in ten chapters plus four appendices, Ushonye Ayim, a non-scientist novelist and cleric, is poised to challenge medical experts on the issues of HIV/AIDS. This book **"HIV/AID-How expert spread the virus"** raises queries based on the scientific statements and knowledge available to him from the medical literature. Of course, it makes interesting reading and is a challenge to health scientists to continue to research, update and disseminate the existing information concerning this scourge. Indeed changes in the characteristics of the human tissues or organs, disease manifestations and new diagnostic tools are being advertised at a very tremendous frequency. Fortunately, confusions and knowledge gaps that have existed are being continually resolved. For instance the link between HIV and leukaemia, which does not really exist, has long been resolved. Also, it's relationship with the African Mackay monkey as the original source of the HIV does not hold anymore.

Basically, certain scientific fundamentals on the disease are obvious. For instance, the difference between Human Immunodeficiency Virus (HIV) and Acquired Immune Deficiency Syndrome (AIDS) needs to be explained in a simple language. The HIV is just an organism or germ which can enter the human body through various means. The means of transmission may involve blood transfusion, sexual intercourse, injections with contaminated needles, breast milk and kissing. The virus has to find a favourable environment in the human body where it can multiply into millions before it can cause a problem. In a large number, it destroys or suppresses the human body immunity or defence mechanism thus rendering the infected victim susceptible to other opportunistic infections from bacteria, fungus, parasites and other viruses. It is when such an opportunistic disease manifests in the victim of HIV that the label 'AIDS' can be applied. HIV infection alone is not directly equivalent to AIDS.

The foetus (unborn child) can easily derived the virus from the mother across the placenta hence he is born with HIV. To agree with the author, this is clearly a congenital infection. However, medical science has not been able to show that such infants are otherwise afflicted by the virus implying that an infected baby does not develop AIDS in the uterus. It is only when the virus has adversely affected the infant's immune system and the infant is exposed to other infections outside the uterus then can he develop AIDS. Since the infant was born with the virus only, and not AIDS, but acquired the AIDS after birth, the nomenclature 'ACQUIRED' is appropriate as is applied in all cases.

A high concentration of HIV is found in the human semen, blood and cervical/vaginal secretions. Except in very ill AIDS patients, the concentration in breast milk, saliva and urine is known to be low.

For the HIV to cause or create any danger to the human, it has to work in a large number which is referred to as a viral load. In persons whose body defence or immunity is strong and powerful, a small dose of the virus which enters such an individual can be suppressed, kept in check and may later get eliminated. Conversely, when a huge quantity of the virus enters the victim and his body defence is not strong enough to fight the virus, it favours virus multiplication within the recipient's body. Clearly, there are records of people who have acquired the virus but have been living in a healthy state with it for years without developing AIDS.

This may also explain why a widow of a man who died of AIDS can herself remain healthy and may not even test HIV-positive (chapter 5). Nevertheless, if the intrinsic body defence had failed to destroy the virus, the latter may get reactivated, multiply and become virulent even after years.

There are many modalities for preventing the spread of HIV or AIDS. The author is right in raising the issue of pores in the latex condom through which the virus can leak during sexual intercourse. But the likelihood of this type of leak NOT occurring during vaginal intercourse is more than 99.9% leaving only a probability of less than 0.1% that the virus can wriggle through. Wearing double condom during performance may totally close this gap. Nevertheless, if by any imagination, there is such a free route, the number of the virus that can come through will be so minimal that the body defence will suppress it so that the recipient of the virus is most unlikely to suffer any harm.

Chapters three and four of the book focus on saliva as a medium for the transmission of the infection through kissing and oral sex. As earlier stated, saliva is known to be one of the body fluids with minimal viral load. By reasoning of this low concentration of HIV in human saliva, the chance of the few viruses surviving to cause a problem is also minimal or almost nonexistent.

The author scores a point by amplifying a known alarm over a possible increase in transmission frequency if the mucosa of the recipient's lips and oral cavity has sores on them as this can enhance the danger of trauma to the penis. There is a tendency for admixture of blood, genital fluids and saliva hence the danger of transmission. It may not be possible to provide a protective sheath for the mouth and the tongue for the for the purpose of oral sex especially for the lesbians who engage in leaking their genitals. Nevertheless, penile condoms for oral and anal sex remain protective.

The practical or theoretical point regarding the transmission of the virus through saliva applies almost directly to breastfeeding. The low viral load in human milk, difficulty in the HIV penetrating the intact oral mucosa and gut wall, and the acid medium of the stomach, all play a positive role in protecting the breastfed infant. Mothers who are capable of providing breast milk substitutes are advised to avoid breastfeeding.

I want to agree with the author that even though this disease has been extensively researched, there still remain a lot of unknown and grey areas. The scientist can only reveal all that he knows about a disease condition at a point in time. He then continues to research so as to augment and update available information.

In human medicine words like NEVER and ALWAYS are very rarely applied to disease conditions. This appears to be the meeting point between this book writer and the scientist on HIV/AIDS. The scientific statements made on the disease are based on clinical observations, experience and laboratory findings. No scientist will categorically claim or pronounce that the virus which is related to AIDS will ALWAYS or NEVER do this or that. All the experts are claiming is on the statistics or chances of the virus being able to behave in one way or the other based on their experience and the data available there are still lots of mysteries surrounding HIV/AIDS which are yet to be unraveled. The challenges posed by Ushonye Ayim should spur the scientist to work harder so that more axiomatic statements can be released to the public.

Since there are still areas of doubt and uncertainties, it becomes mandatory for the generality of persons to be mindful of their sexual behaviour. The key point is that one hundred percent safety cannot be guaranteed from any of the protective measures recommended, and the disease remains incurable for now. Moral and religious sanctity is the answer to keep oneself virus-free. As pastor Ushonye Ayim posits, if people were to live with the fear of the Almighty God, they will be wise enough to keep off avenues that lead to the acquisition of the deadly virus. I want to join the author/pastor in admonishing those with reckless sexual behaviour that the very minute chance through which one acquires the HIV might be yours at any time.

This may follow just one exposure either through kissing or a penetration through a pore in a condom. Anyone around you is a suspect until proven otherwise; a healthy carrier of HIV is most dangerous in disseminating the infection.

The STATPAK, DETERMINE, CAPPILLUS and the Polymerase Chain Reaction (PCR) are diagnostic test kits known to be very sensitive in detecting and confirming the infection in both patients of AIDS and the healthy carriers of the virus. PCR test is now available in Africa. A good number of health facilities do offer free HIV screening but most people are too scared, superstitious or too care free to be bothered about their HIV status.

For me, Ushonye's communication is really not an attack on medical experts but is an eye-opener and a direct warning to the public to keep off danger.

Asindi A. Asindi, MBBS (Ibadan), FRCP (Glasg), FRCP (London), Dip Theo

Professor of Paediatrics

University of Calabar, Nigeria.

POSTSCRIPT II

This 80 page book is an eye-opener to the truths of our present day malady – HIV/AIDS. The Author has driven a wedge through the commonly held creed of **'The Basic Facts of AIDS'** propagated by Experts in the campaign against HIV/AIDS.

These Authorities may dismiss the Author's views with a wave of the hand but make a lot of sense for the common person in the street.

Take the issue of abstinence as promoted by both faith based organization and the secular world. Does its practice portend the fear of God or the fear of HIV/AIDS? Will abstinence be promoted if there was a condom that was 100% efficacious? The Author's view is on the contrary.

In his regard the love of God should be the overwhelming motivation for the practice of abstinence.

The Postulation by the Author that the mosquito being a cure for human ill health and particularly to HIV/AIDS is an interesting one that Scientists should look into. Scientists believe that the AIDS virus cannot survive in the mosquito's stomach because of the acidity in the stomach. But then what about the acidity in vaginal secretions?

All said and done, '**HIV/AIDS: How Experts Spread the Virus'** is recommended for all those who want to live their lives without the fear of AIDS.

Dr. John O. Odok

Chief Medical Officer & Former Programme Manager

Cross River State HIV/AIDS Programme

Ministry of Health

Calabar, Nigeria.

BIBLIOGRAPHY

The Medical and Health Encyclopedia ed. Richard J. Wagman, Associate Clinical Professor of Medicine.

Your Health and You, volume 1. Harold Shryock and Hubert O. Swartout in Collaboration with 38 leading Medical Specialist.

A challenge to Love: Understanding HIV/AIDS, Léonie McSweeney.

Answers To Questions on HIV/AIDS, NYSC Reproductive Health Booklet.

Society for Family Health: Leaflet on HIV/AIDS Prevention.

National Action Committee on AIDS: Fact sheet on HIV/AIDS.

Practical Guide on the Study and Prevention of HIV/AIDS/STDs, Ubana A. Ofor.

Planned Parenthood Federation of Nigeria: Leaflet on HIV/AIDS Prevention.
BBC World Service/International Planned Parenthood Federation: Sexwise Guide.

National AIDS/STD Control Programme Booklet AIDS: What Teachers Should Know.

New Structures of International Communication: The Role of Research, Cees Hamelink.

Host Genes and HIV: The Role of the Chemokine Receptor Gene CCR5, Janet McNicholl. www.cdc.gov

Selection of CC Chemokine Receptor 5-Binding Peptide from a Phage Display Peptide Library, Wang Fang-Yu, et al. www.sciencelinks.jp

Concordance between the CC Chemokine receptor 5 genetic determinants that alter risks of transmission and disease progression in children exposed perinatally to human immunodeficiency virus, Mangano Andrea, et al. www.cat.inist.fr

ABOUT THE AUTHOR

Ushonye Ayim is a Revolutionary Author of african descent,his Postulation on a vaccine that promises a cure against the 'AIDS Virus'is part of his first work published in the United States and read throughout the planet,titled,'HIV/AIDS-How Experts SPread The Virus'.He also wrote some other work which promises to publish.These works include:Poverty of Ideas,Thinker Paradise,The Mechanism of Confidence and Armoury of the Thinker.He is also the 'Ruggedly-Anointed Man of God' under the Radical Gospel Ministry-Reconciliation Mysticism.

He is of the opinion that the greatest asset God gave man is the human mind. He maintains that HIV/AIDS experts have not been able to enlighten the world properly on AIDS. They, many times spread the virus through the message.

He is from Ishiborr in Ogoja Local Government Area of Cross River State of Nigeria.

e-mail: pepperestedpriest@yahoo.com

ABOUT THE BOOK

This book, **HIV/AIDS – HOW EXPERTS SPREAD THE VIRUS**, is indeed an Eye-opener. It is the fastest medium known to mankind in curbing the spread of the dreaded human virus. This text is a product of Scientific Revolution. According to Kuhn, in 'The Structure of Scientific Revolutions (1962) *'A Scientific Revolution is needed if research is to contribute meaningfully to the transition from old structures to new structures'*. For it is intended to provoke the medical scientist to return with spontaneous immediacy to the clinical laboratory to further the search for superior preventive measures against the scourge. This will go a long way to boost the medical profession and heighten its estimation in the minds of the public. This work, which is quite timely, is a rewarding 'rescue effort' against the spread of the AIDS virus from the infected to non – infected. Meanwhile, to justify the fact that 'scientific civilization' supports the principle of 'freedom of thought'.

This text carries on it a very revealing **'Foreword'** from an objective seasoned Medical Personnel.

INDEX

<u>NOTES</u>

<u>NOTES</u>

NOTES

NOTES

NOTES

NOTES

NOTES

NOTES

NOTES

www.ingramcontent.com/pod-product-compliance
Lightning Source LLC
Chambersburg PA
CBHW051448280526
45785CB00003B/1473